Goran Šibenik

Energy saving using smart glazing systems

AF141248

Goran Šibenik

Energy saving using smart glazing systems

Description of thermal building performance simulation and analysis of the results

Natural Sciences Series

Impressum / Imprint

Bibliografische Information der Deutschen Nationalbibliothek: Die Deutsche Nationalbibliothek verzeichnet diese Publikation in der Deutschen Nationalbibliografie; detaillierte bibliografische Daten sind im Internet über http://dnb.d-nb.de abrufbar.
Alle in diesem Buch genannten Marken und Produktnamen unterliegen warenzeichen-, marken- oder patentrechtlichem Schutz bzw. sind Warenzeichen oder eingetragene Warenzeichen der jeweiligen Inhaber. Die Wiedergabe von Marken, Produktnamen, Gebrauchsnamen, Handelsnamen, Warenbezeichnungen u.s.w. in diesem Werk berechtigt auch ohne besondere Kennzeichnung nicht zu der Annahme, dass solche Namen im Sinne der Warenzeichen- und Markenschutzgesetzgebung als frei zu betrachten wären und daher von jedermann benutzt werden dürften.

Bibliographic information published by the Deutsche Nationalbibliothek: The Deutsche Nationalbibliothek lists this publication in the Deutsche Nationalbibliografie; detailed bibliographic data are available in the Internet at http://dnb.d-nb.de.
Any brand names and product names mentioned in this book are subject to trademark, brand or patent protection and are trademarks or registered trademarks of their respective holders. The use of brand names, product names, common names, trade names, product descriptions etc. even without a particular marking in this works is in no way to be construed to mean that such names may be regarded as unrestricted in respect of trademark and brand protection legislation and could thus be used by anyone.

Coverbild / Cover image: www.ingimage.com

Verlag / Publisher:
AV Akademikerverlag
ist ein Imprint der / is a trademark of
OmniScriptum GmbH & Co. KG
Heinrich-Böcking-Str. 6-8, 66121 Saarbrücken, Deutschland / Germany
Email: info@akademikerverlag.de

Herstellung: siehe letzte Seite /
Printed at: see last page
ISBN: 978-3-639-48982-8

ACKNOWLEDGEMENTS

I would like to express my gratitude to my supervisor Univ.Prof. Dipl.-Ing. Dr.techn. Ardeshir Mahdavi for his engagement during the process of writing this Master thesis. With his useful comments and remarks he contributed and guided me to shape this work into its final form.

Furthermore, I would like to thank Projektass. Dipl.-Ing. Mag. rer. soc. oec. Dr. techn. Robert Zach and Univ. Ass. Farhang Tahmasebi, MSc for helping me to overcome problems that occurred during this period, and devoted their time whenever I needed their help. Without their input this work wouldn't have been done with the same pleasure and enthusiasm. I would also like to thank Ms. Elisabeth Finz for all the help with the administrative tasks.

Last but not least important, I owe more than thanks to my family members, my parents, sister and my grandmother, for all the support and encouragement during my studies. Without their help it would have been impossible to finish my studies. I will always be grateful.

ABSTRACT

A lot of effort has been put into improving energy performance of buildings, and still a lot of problems are found with thermal performance of windows. Although its U values are constantly being improved, overheating problems can often occur in buildings with large areas of transparent surfaces. Besides the conventional shading products, some innovative solutions already exist on the market, but their efficiency and performance are still not examined thoroughly. Smart glazing, and especially electrochromic window devices, is one of the most promising products that are commercially available. Its implementation in simulation programs, as well as the results of the simulations, is tested and analyzed in this research. Simulation is performed by coupling lighting simulation program and energy simulation program through the third program – Building Control Virtual Test Bed. Cooling loads of a commercial building are calculated and the possibilities of energy saving by using electrochromic windows are found. Cooling loads are compared to conventional windows, where the effect of electrochromism is examined. Simulations have also been performed with windows with shades. The results demonstrate potential cooling loads reduction when smart glazing is used in oceanic climates.

Keywords: smart glazing, smart windows, transmittance, solar heat gain, electrochromic, energy saving, cooling load, shading, illuminance, Radiance, EnergyPlus, BCVTB, daylight

ABSTRACT

Es wurden bereits viele Anstrengungen unternommen die Energieeffizienz von Gebäuden zu verbessern. Die thermische Leistung von Glasflächen verursacht jedoch noch einige Probleme. Obwohl der „U" - Wert ständig verbessert wird, entstehen oft Überhitzungsprobleme in Gebäuden mit großen Fensterflächen. Zusätzlich zu herkömmlicher Verschattung, sind einige innovative Lösungen verfügbar, deren Effizienz und Leistung jedoch noch nicht ausreichend untersucht wurden. „Smart-Glazing", und vor allem elektrochrome Fenster, ist einer der vielversprechendsten Produkte, die im Handel erhältlichen sind. Seine Umsetzung in Simulationsprogrammen, sowie die Simulationsergebnisse, werden in dieser Studie getestet und analysiert. Die Simulation wird durchgeführt durch die Koppelung eines Programmes zur Beleuchtungssimulation (Radiance) und einem Energiesimulationsprogramm (EnergyPlus). Die Kommunikation zwischen diesen beiden Simulationsprogrammen erfolgt durch die Software „Building Control Virtual Test Bed" (BCVTB). Dabei wird der Energieverbrauch für die Kühlung von herkömmlichen Gebäuden berechnet und mögliche Energieeinsparungen durch den Einsatz von elektrohromen Fenstern eruiert. Zusätzlich wurde diese Analyse für den Einsatz von aktiver Verschattung bei herkömmlichen Fenstern durchgeführt. Die Ergebnisse zeigen die Reduktion des Kühlbedarfs, wenn „Smart-Glazing" in Regionen mit Meeresklima verwendet wird.

Schlüsselwörter: smart Verglasung, intelligente Fenster, Durchlässigkeit, erhitzen durch die Sonne, Elektrohromie, Energie sparen, Energieverbrauch während der Kühlung, Sonnenschutz, Beleuchtung, Radiance, Energy, BCVTB, Tageslicht

CONTENTS

1 INTRODUCTION

1.1 Glazing development

"Window is an opening constructed in a wall or roof that functions to admit light or air to an enclosure and is often framed and spanned with glass mounted to permit opening and closing". (Houghton Mifflin Company, 2006) Traditional windows had a role of providing light in the buildings, fresh air and view to the occupants. Commercial buildings can often be found as completely sealed, mechanically ventilated and electrically lit. Traditional role of windows has been lost in order to provide better energy efficiency of buildings.

However, in order to design in a more human-oriented approach, natural light should not be neglected. Daylight contributes in improving occupants' satisfaction, productivity and health. It is not possible to replace it by electrical lighting. Natural light changes direction, intensity and color during the day, and connects occupants to the outside world. Different types of glazing used in building can behave differently when they transmit daylight. All the glazing materials have property of visible transmittance. It determines how much visible light is entering the room.

Besides the visible light, windows transmit other direct or indirect radiation. In that way heat can be transmitted into the room. "The ability to control this heat gain through windows is characterized in terms of the solar heat gain coefficient (SHGC) or shading coefficient (SC) of the window." (Regents of the University of Minnesota, 2011)

Windows play an important role in the thermal comfort of occupants. U value represents overall heat-transfer rate or the insulation of the window. When there is temperature difference between inside and outside, then there is an energy loss by convection, conduction and radiation. Windows can't reach heat-transfer rate of other building elements. That is why windows are colder surfaces in the room during the winter even though room temperature can be high enough. People loose energy by radiation more as they get closer to the window.

Relevant since	Floor to ground	Top Floor Ceiling	External Wall	Roof	Window	g-value	Door (External)
15.11.1976	0,85	0,71	1,00	0,71	2,5	0,67	2,5
01.10.1993	0,4	0,20	0,5	0,2	1,9	0,67	1,9
26.10.2001	0,45	0,25	0,5	0,25	1,9	0,67	1,9
12.07.2008.	0,4	0,2	0,35	0,2	1,4	0,67	1,7
01.01.2013.	0,4	0,2	0,35	0,2	1,4	0,67	1,7

Table 1:   Overview of required U-values of building elements
through time in Vienna (Österreichisches Institut für
Bautechnik 2007, 2011)

Another thermal comfort issue is drafts. Cold air can enter through leaky windows, or windows cool down the air next to it, so warmer air from the ceiling replaces it, making the room loose more heat.   These thermal comfort issues represent bigger problem during the cold than during the warm weather.

Direct sun represents a problem to the occupants. Glare is produced by too much daylight. It causes difficulties to see due to differences between areas lit by entering sunlight and the area that an occupant is trying to see. Glare within the range that eye can handle is called discomfort glare, and the glare that eye can't handle is disability glare. When person views source of illumination it causes direct glare. Indirect glare is caused by the reflected light. People put the blinds on even though it means they can't have the view to the outside, and that they will have to use electrical lighting.

Some of the window properties can be useful on certain environmental conditions, and present a problem during other days. For example, when poorly insulated tinted windows are used, window temperatures can be higher than with the clear glazing. During the cold sunny days these windows can get warmer than the clear ones. But when it is a hot sunny day, they can get too hot, and affect thermal comfort of people.

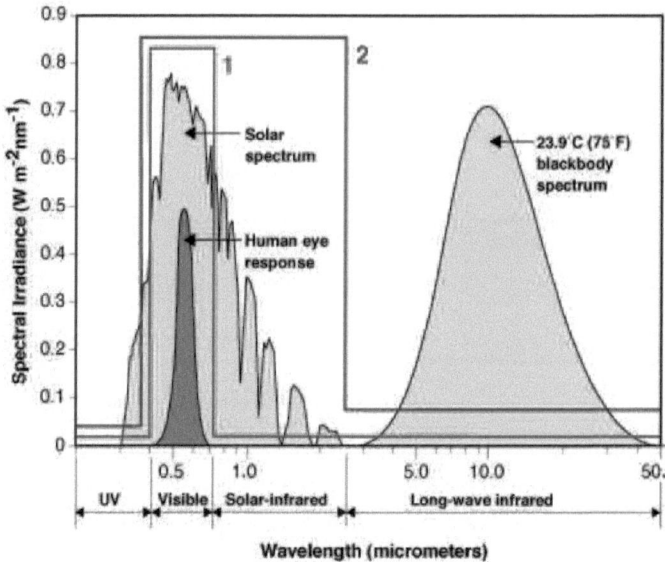

Figure 1   **Ideal spectral transmittance for glazings in different
climates (McCluney, 1996).**

1. Idealized transmittance of a glazing with a low-E coating designed for low solar heat gain. Visible light is transmitted and solar-infrared radiation is reflected. Long-wave infrared radiation is reflected back into the interior. Aim of this approach is to reduce solar heat gain and this approach is suitable in almost all climates.

2. Idealized transmittance of a glazing with a low-E coating designed for high solar heat gain. Visible light and solar-infrared radiation are transmitted. Long-wave infrared radiation is reflected back in the interior. This approach is more commonly used in cold climates where solar gain is wanted (McCluney, 1996).

1.2   Smart glazing

Smart materials are defined differently, based on subjects that define them. NASA defines them as 'materials that "remember" configurations and can conform to them when given a specific stimulus' (NASA). In Encyclopedia of Chemical Technology: 'smart materials and structures are those objects that sense environmental events, process that sensory information, and then act on the environment' (Kroschwitz 1992). "Whether it is a molecule, a material, a composite, an assembly, or a system, 'smart materials and technologies' will exhibit the following characteristics:

- Immediacy – they respond in real-time.
- Transiency – they respond to more than one environmental state.
- Self-actuation – intelligence is internal to rather than external to the 'material'.
- Selectivity – their response is discrete and predictable.
- Directness – the response is local to the 'activating' event" (Addington, Schodek 2005)

There are various different techniques applied in smart window products. The most important rule which has to be obeyed is that a transparent mode has to be possible. There are currently three technologies with external triggering signal that are used for smart windows: chromic materials, liquid crystals and electrophoretic or suspended-particle devices. Chromic devices can be photochromic and thermochromic, which change automatically by environmental conditions (light and temperature), and electrochromic and gasochromic.

Most important aspect of smart windows is their transmittance in visual and whole solar spectrum modulation range, and the focus of this work is set to this property. Besides that, very important aspect is expected lifetime and number of achieved cycles. Switching time, achieved window size, energy consumption, operating voltage and operating temperature range are also important for the evaluation of smart windows. (Baetens et al. 2010, 87-105)

System type	Spectral response (bleached to colored)	Interior result visual	Interior result thermal	Input (energy)
Photochromic	Specular to specular transmission at high UV levels	Reduction in intensity but still transparent	Reduction in transmitted radiation	UV radiation
Thermochromic	Specular to specular transmission at high IR levels	Reduction in intensity but still transparent	Reduction in transmitted radiation	Heat (high surface temperature)
Electrochromic	Specular to specular transmission toward short wavelength region (blue)	Reduction in intensity	Proportional reduction in transmitted radiation	Voltage or current pulse
Gasochromic	Specular to specular transmission toward short wavelength region (blue)	Reduction in intensity	Proportional reduction in transmitted radiation	Gas
Liquid crystal	Specular to diffuse transmission	Minimal reduction in intensity, reduction in visibility, becomes diffuse	Minimal impact on transmitted radiation	Voltage
Suspended particle	Specular to diffuse transmission	Reduction in intensity and visibility, becomes diffuse	Minimal impact on transmitted radiation	Current

Table 2   Smart Windows, (edited, original Addington, Schodek, 2005)

After the existing technologies have been examined, two commercially available smart window technologies have been taken into further consideration for the utilization in façade of commercial building. First type is the electrochromic window and the second one is a suspended-particle smart window (SPS window). Besides the two mentioned ones, a liquid crystal smart windows are also commercially available. Because it needs high amount of input energy for the transparent mode, implementation in the commercial building would have much larger energy demands if daylight was used (building is in transparent state most of the time). This type of window has not been considered in the rest of this research.

Several companies offer products based on electrochromic and SPS window technologies. There are only three companies that produce electrochromic windows for exterior building glazing currently on the market: SAGE Electrochromics, EControl-Glas and Gesimat. Research Frontiers Inc. is the licensor of SPD technology. Not enough details and data about the application of this technology in the building exterior, or about the properties of these windows were found. SPD technology also has higher power consumption because it needs constant power for maintaining the clear state. It changes between clear and bleached states faster than other smart window products. This technology is already widely present in automotive and aircraft industry. It has also good properties for implementation in buildings. For the simulation purposes, electrochromic window properties are considered. Other technologies of smart glazing work in different ways, but have similar final effects, so results of this research might also prove their advantages.

Figure 2   E-control glazing in meeting room in Plauen (Econtrol official website)

After the analysis, EControl standard double IGU was chosen as the smart window device for the simulation. It is an electrochromic window that uses tungsten oxide as the electrochromic active layer. It can be controlled automatically or manually. It is already used for the building fenestration and requires electrical power only for changing the transmission state. One complete switching cycle takes less than 0,5 Wh/m^2, and one controller consumption during switching is less than 10 W. The standard DGU is 29 mm thick, consisting of 9 mm electrochromic pane (outer surface), 16mm cavity and 4mm inner pane with low-E coating.

1.3   Electrochromic windows

Ability of a device to change optical properties when external potential is applied is called electrochromism. It is related to the processes of ion insertion and extraction. These devices consist of several layers. Basis of electrochromic device is glass or plastic. There is electrochromic coating deposited on it. "Electrochromic coatings (EC) are switchable thin-film coatings applied to glass or plastic that can change appearance reversibly from a clear to a dark Prussian Blue tint when a small DC voltage is applied" (Lawrence Berkley National Laboratory, 2006) Prussian Blue appears only in certain coatings, but also in the "EControl standard double IGU". Electrochromic coating is one nanometer thick and it consists of several layers. Its first layer is a transparent conductor, to which cathodic electroactive layer (or more of them) is attached. They are followed by ion conducting layer and ion-storage film or one (or more) anodic electroactive layer(s). It ends with another transparent conducting film.

It works as a battery. When bipolar voltage is applied, lithium ions move from ion-storage film through electrolyte to electroactive layer. A reverse electrochemical reaction takes place and causes tinted appearance. With the reverse voltage applied, ions move back and the device gets into the bleached state again.

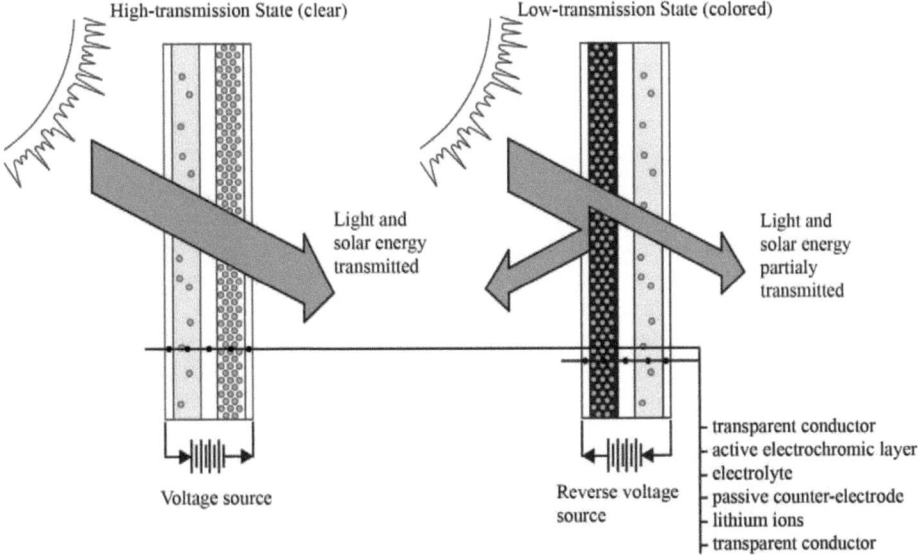

Figure 3   Diagram of a typical tungsten-oxide electrochromic coating (redrawn from Lawrence Berkley National Laboratory, 2006)

Tungsten oxide is currently the most used electrochromic material and it is used also in the "EControl standard double IGU" which is examined in the research. Its behavior as active electrochromic layer can be represented by the chemical reaction

(transparent state) $WO_3 + xM^+ + xe^- \leftrightarrow M_xWO_3$ (deep blue)

13

Where M^+ is a molecular ion that can be H^+, Li^+, Na^+ or K^+, $0 < x < 1$ and e^- are electrons. There are other electrochromic metal oxides besides tungsten oxide like nickel oxide, iridium oxide, niobium oxide, as well as some other inorganic electrochromics and polymer electrochromics, but they have not been studied in this research.

1.4 Motivation

Rising standards for energy performance of buildings encourage implementation of new and innovative materials. In Austria in 2011, energy used for heating, cooling and warm water was 30,34% and for lights and computer equipment 2,95% of total energy consumption (Statistik Austria 2013). Windows are often found as weak point of the energy performance of building, its peak demand and environmental consequences. Although they have developed significantly in the past few years, there are still some problems. Building envelopes are becoming more insulated and that results in lower heating demands. Along with it, windows also become more insulated, and contribute in keeping the heat inside the building. But during the cooling period there are problems with overheating and high cooling loads, even with well insulated windows. Glazing surfaces allow sun energy to get inside the room by radiation, what increases energy necessary for cooling.

Smart windows are a promising solution for the problem. Energy savings in hot climates has already been proven (26% of lighting energy and 20% of peak cooling loads in hot climates as California, USA) (Baetens et al. 2010. 87-105). But there are still not enough data about the advantages of these windows in colder and continental climates.

Besides that, windows represent a dominant role in exterior and interior perception of building. Smart windows provide view preservation, as other physical obstacles like blinds or shades are less used or not used at all. There is also no need for additional elements on the façade to prevent sunlight, if it can be controlled by glazing itself. This can produce "cleaner" geometries which is a design trend of modern commercial buildings.

Smart windows are already used in buildings, but so far, it is not a technology that is supported by common energy performance simulation programs. One of the aims of this work is to find currently possible ways of simulating performance of this technology in simulation programs. Intention is also to examine the complexity of the whole process, and necessary knowledge for performing the simulation.

Building automation systems already control window shading and HVAC systems in buildings, so from that aspect it would be possible to implement a system with smart glazing into the building. Smart glazing can work and be operated by like other shading devices. Combined systems of smart glazing and other shading systems might be more complicated in order to provide optimal conditions for occupants. These systems have not been analyzed with all their possibilities within this work.

Currently these windows are not commonly used for many reasons. One of the reasons is lack of knowledge about their properties, of their advantages and of their applicability. Predictions of the future of world climate say that general world temperature will increase because of the global warming. For that reason cooling loads will increase and heating loads decrease. This should attract more investment and research in improving products that can reduce cooling loads. With wider use and production of smart windows, its price should be reduced and they should become more available.

These awaiting developments are just some of the reasons why this technology has great potential in the future. There are also other fields related to the use and production of these devices like standardization and quality control, simulation programs and building automation systems. In order to provide free use of smart glazing systems they all need to accompany its development.

2 METHODOLOGY

2.1 Simulation programs

Several programs were considered in order to conduct the research. First problem was to adequately represent behavior of smart glazing in simulation program. Few simulation programs have been examined and checked if they can be used for this purpose. Since the focus was set on the daylight analysis, two programs were considered for lighting simulation: Daysim and Radiance. These programs do not have possibility to perform simulation of energy performance of building. Because of that, other simulation programs were examined for energy analysis.

It was necessary that the energy performance simulation program also responds to the changeable properties of glazing. Two options were taken into consideration: EnergyPlus and EDSL Tas. Energy performance tool had to be able to take input from lighting simulation program. After researching possible relations of two different programs, one of them was found as the most appropriate for performing the simulation. It was created for coupling different simulation programs for co-simulations. It is called Building Control Virtual Test Bed (BCVTB). BCVTB is based on Ptolemy II software environment, whose core is collection of Java classes and packages.

BCVTB is linked to the EnergyPlus program for whole building energy simulation and to Radiance for lighting analysis. Because of that EnergyPlus and Radiance were more thoroughly analyzed. Radiance is a set of programs that uses text-based files for lighting analysis calculations. BCVTB calls a program (in the example case C shell script was used) which can be created to manipulate textual

16

files of Radiance. As glazing properties were defined also as a text file, with BCVTB it is found possible to change it during the simulation. EnergyPlus can communicate in several ways with BCVTB. There are three EnergyPlus input objects that can be mapped by external interface: ExternalInterface:Schedule, ExternalInterface:Acutator and ExternalInterface:Variable. Besides that one, two second mentioned EnergyPlus objects can be used as EnergyManagmentSystem variables.

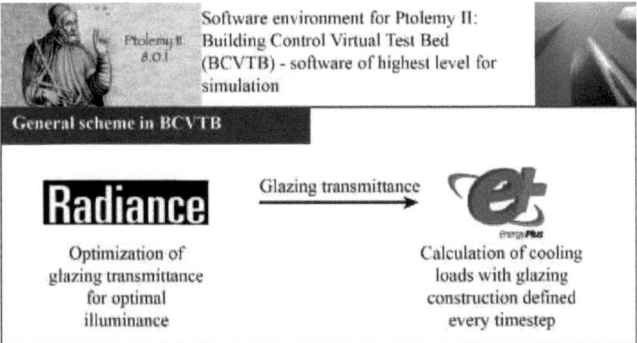

Figure 4   General scheme of simulation programs interaction

EnergyPlus alone does not have an option of changing material properties during its execution so behavior of electrochromic device could be simulated. However, there is a possibility of changing constructions with the help of EnergyManagementSystem. Based on the example that uses thermochromic window it was found that material properties can be changed using EnergyManagementSystem by changing window construction with every timestep.

EnergyManagementSystem (EMS) provides high-level, supervisory control to override selected aspects of EnergyPlus modeling. A small programming language called EnergyPlus Runtime Language (Erl) is used to describe the control algorithms. EnergyPlus interprets and executes Erl program as the model is being run. With the analysis of the selected tools, it was concluded that it is possible to pursue the simulation with these programs and their properties. (The Board of Trustees of the University of Illinois and the Regents of the University of California 2013)

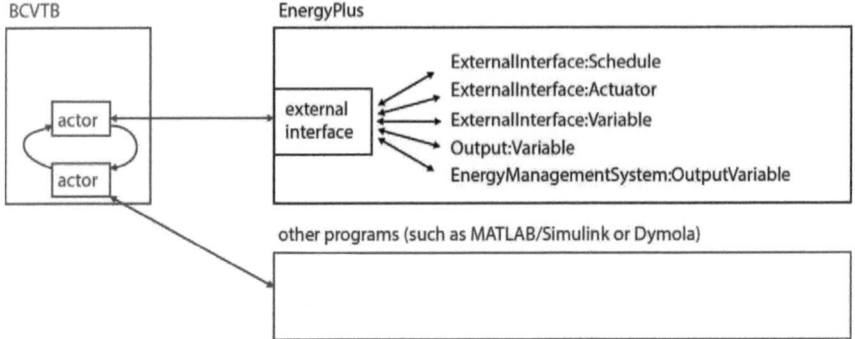

Figure 5 **Architecture of the BCVTB with the EnergyPlus**
client (black) and other clients (grey) (The Board of
Trustees of the University of Illinois and the Regents
of the University of California 2013)

2.2 Light analysis

Lighting analysis is performed with the help of Radiance – a sophisticated lighting visualization system. It is an open-source software and consists of about 100 different programs. The fact that it was primarily created for Unix was one of the reasons why the simulation system is created on Linux operating system. Building was exported from 3d modeling program SketchUp using SketchUp to Radiance Exporter Su2Rad. After the geometry was defined for use in Radiance, all the other data necessary for the simulation was defined. Materials used for the simulation were defined with the help of color picker found on http://www.jaloxa.eu/resources/radiance/colour_picker.shtml. Glazing material, whose properties were changed during the simulation, was created in a separate textual file. (Jacobs 2012)

Radiance is started from BCVTB, with a program written in C shell scripting. First step, before the program starts executing is reading data from weather file. Weather file used is an *.epw file for Schwechat. After that the script is called that finds an optimal transmittance based on the illuminance of measurement point. Transmittance value is exported through BCVTB to EnergyPlus.

The script first creates a sky on every timestep using data from the weather file. Then it calculates optimal value for transmittance of glazing. Optimal value presents lowest transmittance of glazing that provides satisfying illuminance in the measurement point. If there is light (during the day) it first checks if that value is reached with highest or lowest value of glazing transmittance to speed up calculations (because that is the most often case). If not then it finds the optimal value between highest and lowest transmittance. When it is determined, and the illuminance is calculated, both values are exported to BCVTB. The algorithm described is represented in details in next diagrams.

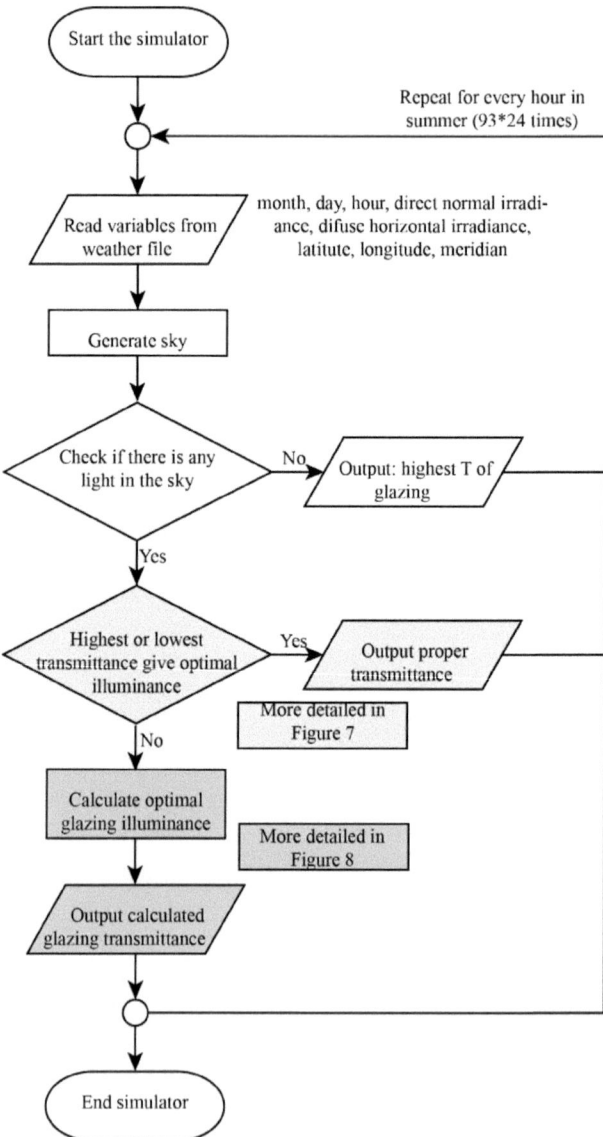

Figure 6 **General concept of transmittance calculating
scenario in radiance**

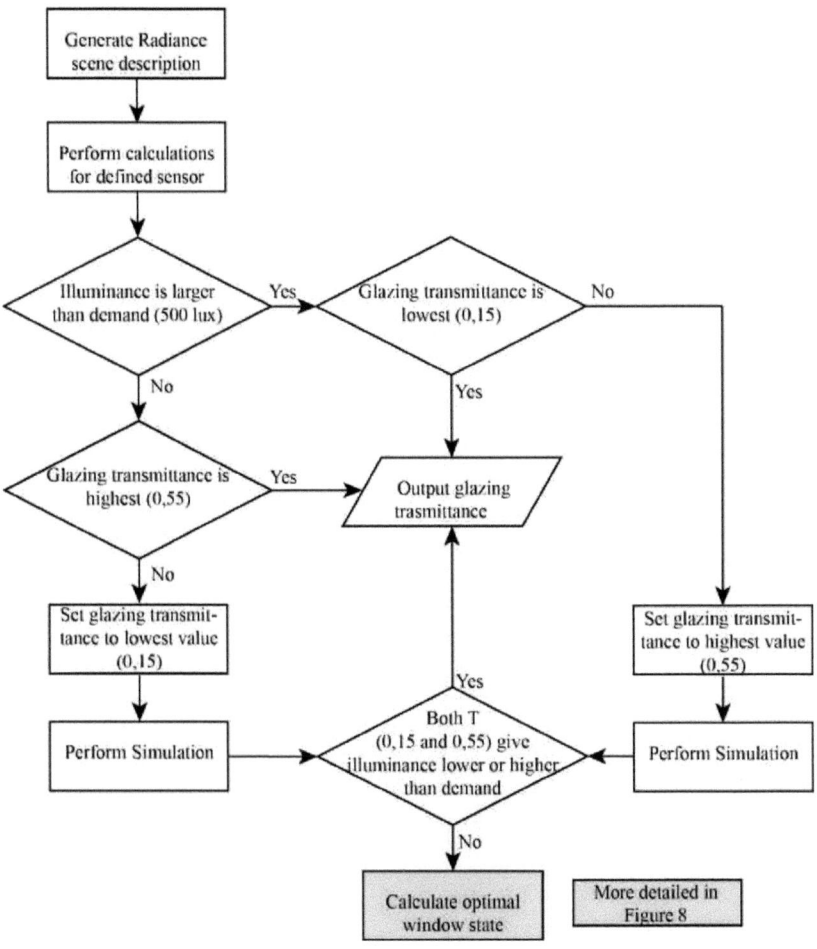

Figure 7   **Checking if minimum or maximum transmittance
gives optimal results**

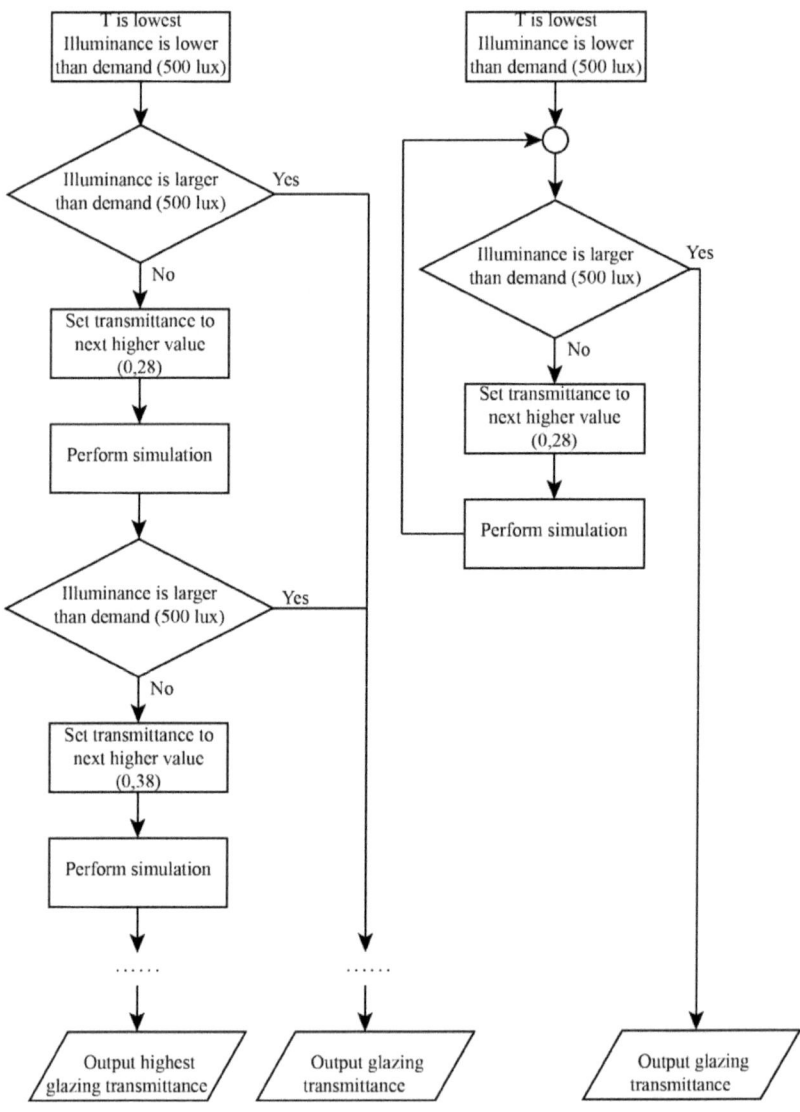

Figure 8   Finding optimal glazing transmittance

Based on the way in which window changes its transmittance coefficients, two ways of finding the optimal transmittance values have been considered. One is used when there is a constant transmittance interval between two neighboring states of smart glazing device. Another approach, like in the researched example, is used for windows that have few modes in which they operate, and there is no logical connection between their transmittance values.

Sensor position is determined after few illuminance calculations in a grid on the working plane have been done. Sensor is oriented towards the ceiling and placed on coordinates that were calculated as the place of median illuminance. Reason for that is because it was necessary to determine the place where the representative daylight illuminance value in the building can be calculated. In the examined cases about 60% of calculated values are between 60% and 200% of the median value. Because of that, it is assumed that if the optimal illuminance is made in the median 500 lux, most of the room will have adequate illuminance (300 – 1000 lux) and electrical lighting will not be necessary.

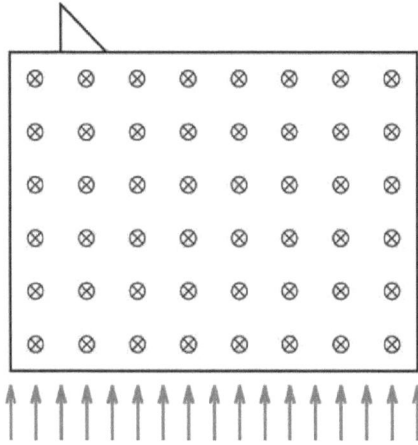

Figure 9   Sensors grid in the building with glazed side
specified

2.3   Thermal performance analysis

When window visual transmittance coefficient is calculated with Radiance, it is exported to EnergyPlus. Energy plus input object is ExternalReference:Variable. As that is an EnergyPlus global variable it can also be used in EnergyManagmentSystem (EMS). In EMS, a program is made that changes construction of window based on the input value. First an EMS:Actuator is created which affects construction of window. EMS program is written in EnergyPlus Runtime Language (Erl). By reading different imported values of imported variable it assigns different constructions for certain building element. There are 5 constructions used, one for every state of the glazing. Construction properties are defined with the help of program Window7 that has some elements of the window already defined and some are calculated and adjusted. External value (glazing visible transmittance) is imported every hour, and simulation timestep is also one hour long.

Thermal performance analysis is a cooling load analysis for summer period, 21 June to 23 September. Windows constructions are defined by value of the visible transmittance every timestep and based on that they change. Results have been compared with the results of the simulation with conventional windows. These windows have exactly the same properties as the electrochromic ones, except they do not change visual nor solar heat gain (transmittance has always highest value). In this way it is possible to value property of electrochromism in energy performance simulation.

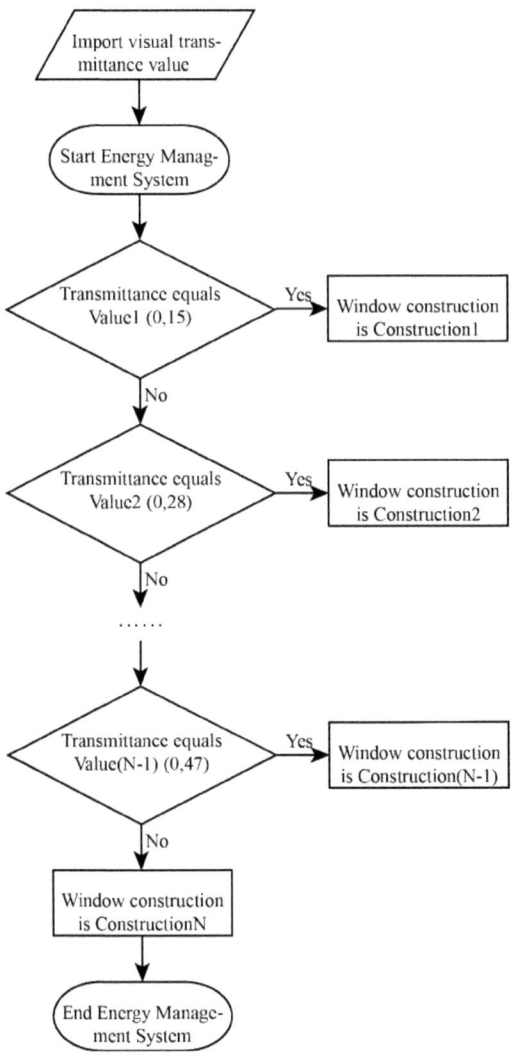

Figure 10   Diagram of Energy plus EMS program

2.4   Connection between simulation programs

The Building Controls Virtual Test Bed (BCVTB) is a software
environment that allows expert users to couple different simulation programs for
co-simulation, and to couple simulation programs with actual hardware. It is based
on Ptolemy II environment. (BCVTB official website)

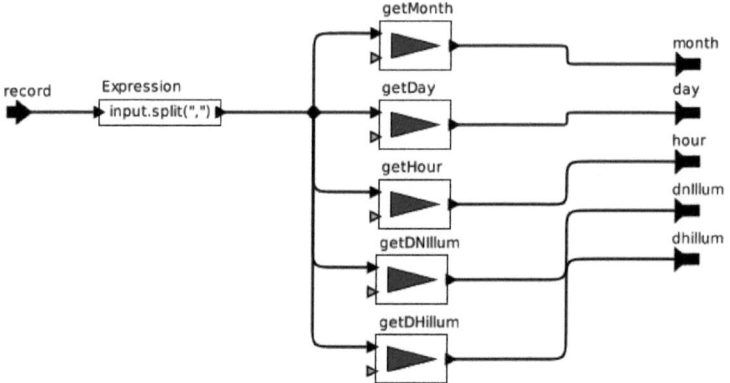

**Figure 11   Data extraction from weather file, BCVTB
screenshot**

There are several steps of Radiance calculations in BCVTB. First data is
read from weather file, and it is done in BCVTB. In weather file every line
contains values for specific hour, for the whole year. First, line for certain hour is
read, then it is converted to array, and specific values are then taken individually
(month, day, hour, direct normal irradiation and diffuse horizontal irradiation).
Starting hour in this simulation is taken by skipping first 8 lines of weather file
and then number of days that are not considered multiplied by 24. These values,
together with latitude, longitude and meridian (they are defined in BCVTB) are
sent to SystemCommand actor where system command is fired with the
arguments. In other words, program that performs Radiance calculation is
executed with imported values. Its performance has been described in light

analysis chapter. Its output is optimal window transmittance, and shades state in latter calculation.

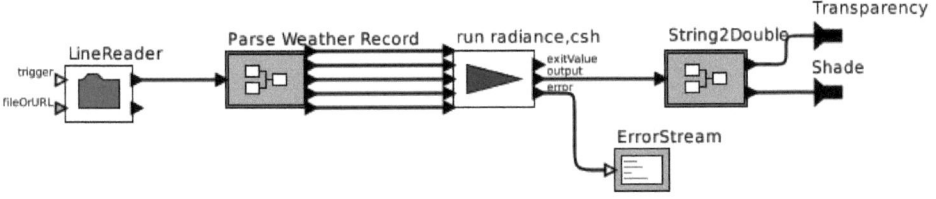

Figure 12   Radiance simulation setup, BCVTB screenshot

Data type of Radiance output values is defined inside the program. In this case they are string type, which has to be converted to a data type readable by EnergyPlus. Data type conversion string to double first deletes empty spaces in the string, then separates it and takes individual strings. After that they can be converted to other data type (double) for further calculations.

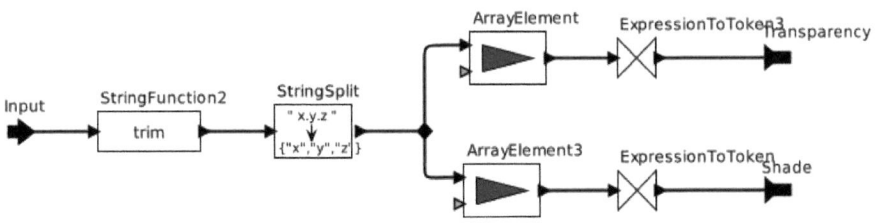

**Figure 13   Data conversion of Radiance output, BCVTB
screenshot**

In the end, data is coupled and sent to EnergyPlus (as array). EnergyPlus is called by Simulator actor which starts data exchange on every firing. Model of computation is defined by Synchronous Dataflow director, which is called SDF director. In SDF domain, execution order of actors is statically determined before the execution. General scheme shows the connection between Radiance and EnergyPlus in BCVTB (Figure 13).

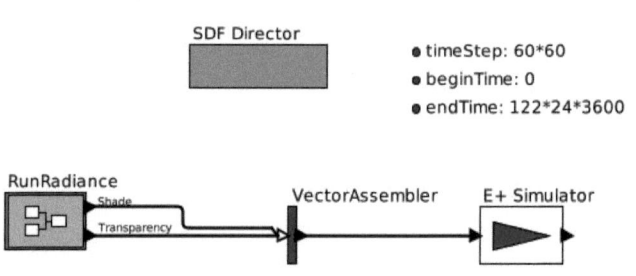

Figure 14   **General scheme of Radiance and EnergyPlus
simulation, BCVTB screenshot**

2.5 Building definition

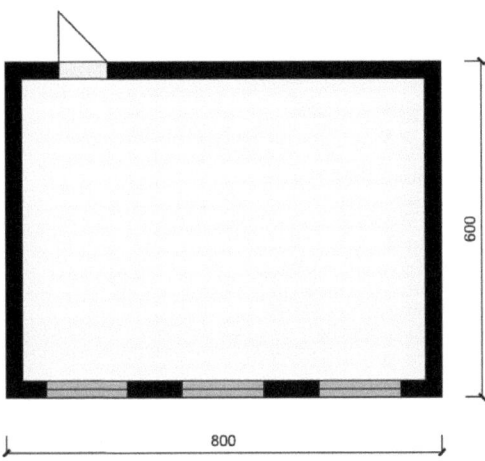

Figure 15   Building floorplan

Simulation building is not an existing real world building. It is placed in the suburban area of Vienna. To simplify the calculation, there are no surrounding objects and building geometry is as simple as possible. Building elements are grouped and have been distinct only by their use in the house (external and internal walls, floor, roof, doors). Floor plan of a building is a rectangular room that is 8 m long and 6 m wide. It is 2,8 m high. Only one longer side of the building is glazed.

Main focus of the building design is windows. There are four main different cases in which the glazed side of the building is oriented to 4 cardinal directions. Besides that, for every orientation 3 sizes of glazing are examined. All the other properties of the building remain the same as cases change.

Construction (Baubook name)	Material	Thickness (cm)	λ (W/mK)
Floor to ground (EFo 01 a)	Parquet	1	0,15
	Cement screed	5	1,7
	Polyurethane foam	1	0,042
	Polystyrene XPS	24	0,041
	Aluminum-bitumen sealing	0,4	0,23
	Reinforced concrete	15	25
	Building paper	0,03	0,17
	Gravel	15	-
	PP fleece	0,02	0,22
Exterior wall (AWm 05 a)	Lime-cement plaster	1,5	1
	Perforated porous brick	25	0,25
	Polystyrene EPS	30	0,04
	Reinforces silicate plaster	0,19	0,8
Roof (DAm 03 a)	Gravel	6	0,7
	Polymer bitumen sealing	0,78	0,23
	PE folium	0,16	0,5
	Polystyrene EPS	36	0,038
	Aluminum-bitumen sealing	0,14	0,23
	PE folium	0,18	0,5
	Reinforced concrete	20	2,5
	Gypsum filler	0,3	0,8

Table 1   Constructions used

Constructions used in building are taken from a passive house catalogue (IBO - Österreichisches Institut für Baubiologie und –ökologie. 2009). Material properties are taken from Die Bauphysik-Datenbank (Kurt Batistti 2008-2013) – mostly from catalogue Basiskatalog Österreich.

2.6   Glazing in building

For the simulation purposes it was of highest importance to determine which glazing is going to be used, and also in which way. Focus was set on commercially available products, with real and measured properties. Smart glazing industry is improving in time and continuously upgrading performances of their products. Visible transmittance and solar heat gain coefficient are being changed, and there is already some spectrum of values available to provide better control of energy and light transmittance. There are several companies that already produce smart glazing suitable for use in buildings. As it was mentioned before, those are electrochromic windows and SPD based glazing. There are several companies that produce electrochromic glazing and the most famous ones are SAGE electrochromics, EControl-glas and Gesimat.

Glazing data was examined in several ways. Some of it was taken from the official websites of the companies, some by direct correspondence with official contact persons in the companies and some was taken from the International Glazing Database (IGDB). IGDB is a collection of measured optical data for many specular glazing products such as uncoated and coated glass, laminates and applied films. Chosen "EControl standard double IGU" was created with help of all three methods. (International Glazing Database official website)

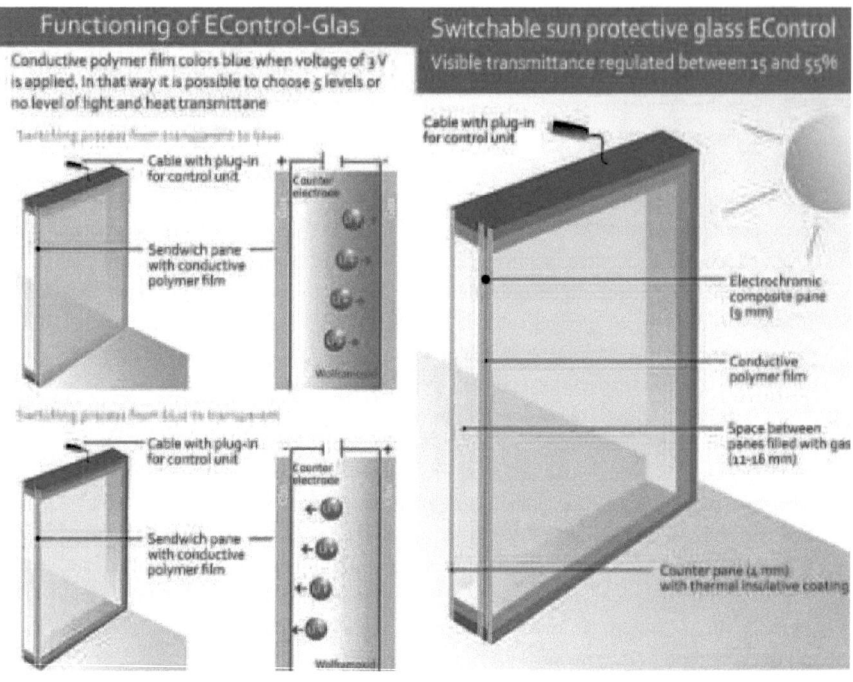

Figure 16 **Technical data of product (translated, original Econtrol official website)**

Window7 software was used to access IGDB, and to model window device that was researched. Properties that were not found remained the same in all the states of the electrochromic device. Properties of low-e glass pane used in this window is already defined IGDB. Properties of the other part of the window are set so they represent main properties of the real product.

State	darkened				bleached
Visible Transmittance	15	28	38	47	55
SHGC	12	21	28	35	40

Table 2 **Transmittance of EControl standard double IGU**

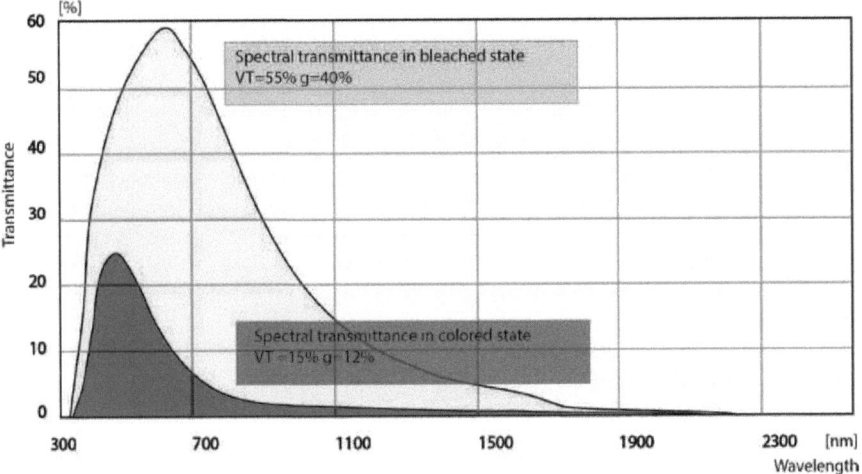

**Figure 17   Spectral transmittance of glazing in bleached and
dark state (Econtrol official website)**

Researched windows are compared to conventional windows, which have same properties as the electrochromic ones, just without the property of changing their visual transmittance and solar heat gain coefficient They have same properties as the electrochromic windows in the state of highest visual transmittance.

2.7   Cases overview

The main difference of electrochromic windows and the conventional windows is dynamic reduction of solar heat gain coefficient. There are several properties of building that could have been examined and compared. Materials and constructions of other building elements have remained constant, with low conductivity in all cases. Only commercial building has been considered because it is the type of building where electrochromic windows are currently mostly used.

Room size hasn't been changed, but window sizes have, so the ratio of glazed area to floor area and glazed area to wall area.

First and main classification of cases was based on the cardinal direction orientation of glazing. For every orientation, 3 window sizes have been examined. First case Win1 has 3 windows, 2,1 m² each, which is a minimum glazing to floor area ratio in commercial buildings -13%, and 28% of wall area. Case Win2 has one window with area 10,5 m², and that makes 22% of floor and 47% of wall area. Case Win3 is with 15 m² of glazed area, which is 31% of floor and 67% of wall area.

Window/Floor Window size Cardinal direction	13 % 6,3 m² Win1	22 % 10,5 m² Win2	31 % 15 m² Win3
South	SouthWin1	SouthWin2	SouthWin3
North	NorthWin1	NorthWin2	NorthWin3
West	WestWin1	WestWin2	WestWin3
East	EastWin1	EastWin2	EastWin3

Table 3   **Cases overview - base calculations**

Three main cardinal directions and 3 windows sizes give 12 examined cases. All of them are compared with corresponding conventional glazing, with the same cardinal direction and window size. Conventional glazing cases give another 12 cases, 24 in total. Their results have determined further examinations.

For shading calculations, focus is set on the window with biggest difference in electrochromic and conventional glazing cooling loads, which was found in case with largest windows. Besides that, all 4 cardinal directions have been considered. Exterior and interior shades have been used and compared to the case without shading devices. That gives another 16 cases, 8 for electrochromic and 8 for conventional glazing.

Shade Cardinal direction	ExtShade	IntShade	NoShade
South	SouthExtShade	SouthIntShade	SouthNoShade
North	NorthExtShade	NorthIntShade	NorthNoShade
West	WestExtShade	WestIntShade	WestNoShade
East	EastExtShade	EastIntShade	EastNoShade

Table 4   **Cases overview - shading calculations**

3 RESULTS & ANALYSIS

3.1 Illuminance – base calculations

Light calculations have been performed first in the coupled simulation, so those results are presented and analyzed first. Based on the electrochromic window light calculation algorithm, calculations of illumination in the room have been performed, and window coefficients optimized. Electrochromic windows showed quite different behavior based on the size of the windows. For the demonstration, illuminance values of building with windows oriented towards south have been shown in Figure 17.

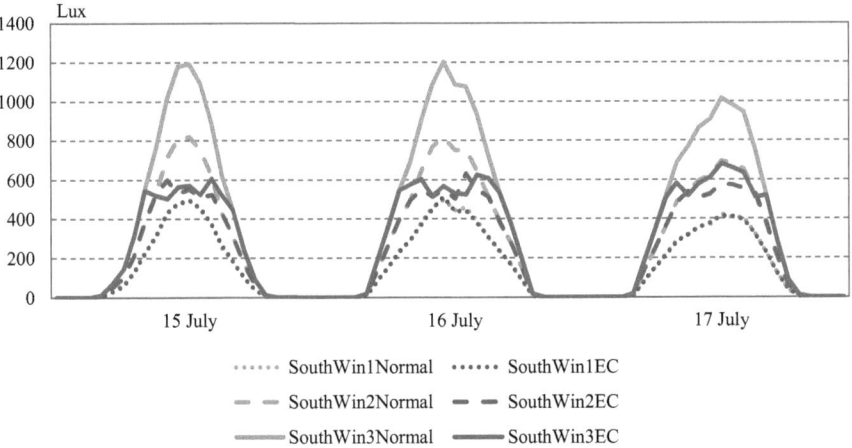

**Figure 17   Illuminance in measurement point - SouthWin1,
SouthWin2 and SouthWin3 - Normal and EC
glazing**

Depending on window size and orientation of building, property of transmittance reduction of electrochromic window is used differently. There are more differences in illuminance calculations results of electrochromic glazing and conventional glazing in building with largest area of transparent surfaces Win3 than in Win1 and Win2. In building Win1 where glazing area is the smallest graphs mostly overlap. Behavior is similar when building is oriented differently, only with peak in illuminance values (during one day) moved to the left or right depending if glazing is oriented to east or west.

Behavior of electrochromic windows is shown on building SouthWin2. When room illuminance in sensor point is too high, window visual transmittance is decreased so the illuminance doesn't go under 500 lux. In these represented days, window decreases its transmittance to first and second lower state of visual transmittance, but it doesn't get to minimum transmittance value.

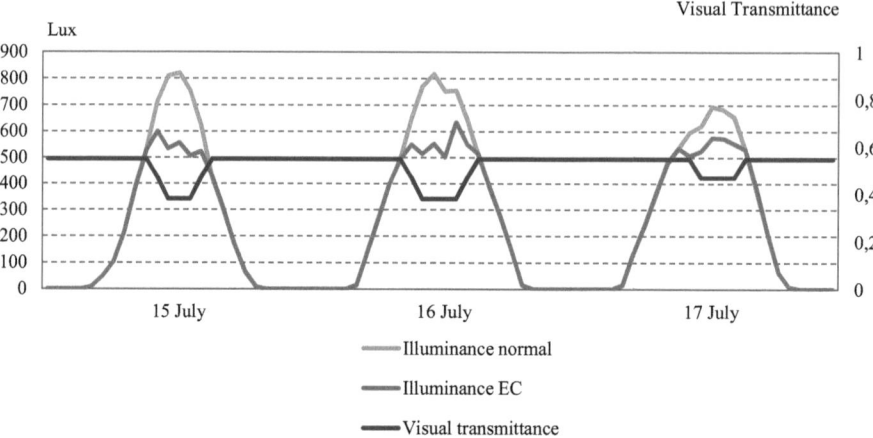

**Figure 18   Illuminance in measurement point - SouthWin2 EC
glazing with Visual transmittance and SouthWin2
Normal glazing**

The problem found in the cases of windows without shading is too high illuminance inside the room. If 750 lux on the place of median illuminance is considered upper border that provides comfortable conditions, where most of the working area will be lighted with the provided daylight, then it is possible to compare number of hours where illuminance is too high.

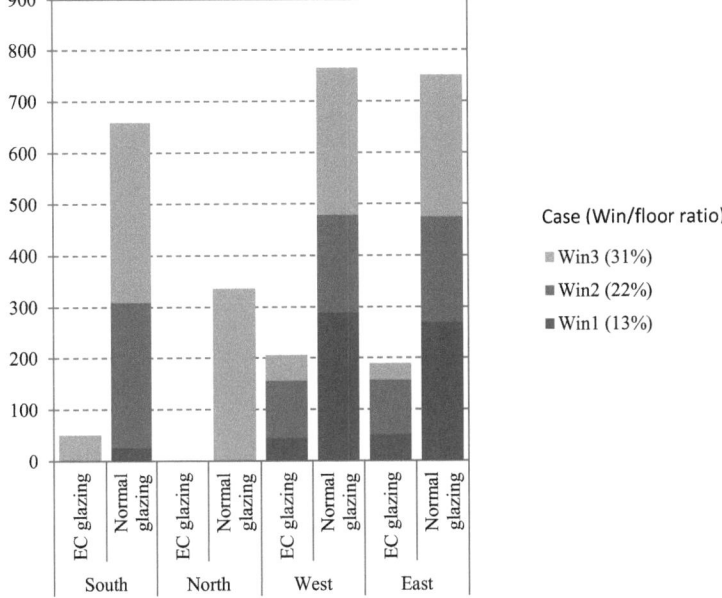

Figure 18   **Number of hours with illuminance larger than 750
lux - base calculations**

This graph shows significant reduction of hours with a too high illuminance when EC glazing is used. In case with smallest ratio of windows and floor areas – Win1, only in East and West cases it is possible to perceive high number of hours with high illuminance. Win2 cases have problems with too high illuminance also in South, and largest windows show problems in all cases. With electrochromic windows too high illuminance is drastically reduced in all the cases where it happens with normal windows. In cases of east and west, there are some problems found even when electrochromic glazing reduces its transmittance values. In north there are no problems with too high illuminance and in south they are insignificant when electrochromic windows are used.

3.2   Energy – base calculations

When base calculations have been performed, it was possible to see relations of electrochromic glazing and conventional glazing, without any shading devices. This graph shows daily differences in cooling loads between buildings with electrochromic and buildings with appropriate conventional window. Differences increase with the size of the window radically. Besides that, east, west and south show bigger differences than north. South shows lower cooling loads differences in the first half, and then it increases (compared to West and East). That is expected because sun gets lower on the sky, so direct illuminance gets deeper into the room in the second part of summer.

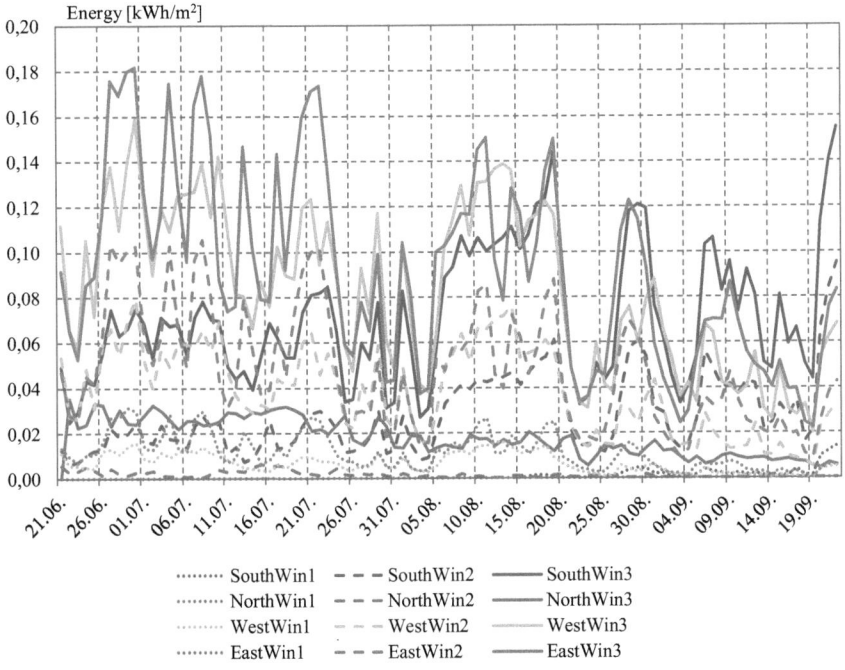

Figure 19   **Difference in cooling loads per m² during summer -
all cases of base calculations**

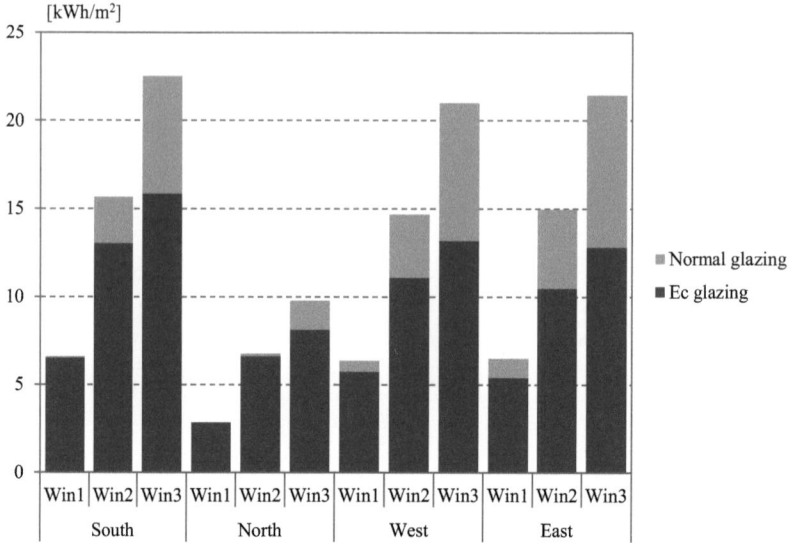

**Figure 20   Cooling loads per m² during summer- base
calculations**

On Figure 20 it is possible to see big differences between cooling loads for
cases with different window sizes. All these differences are reduced with the use
of electrochromic windows. Difference in cooling loads between buildings with
and without electrochromic windows is not same for all orientations. In North, in
case Win3 savings are 17%. Biggest savings are in EastWin3 where cooling loads
are reduced by 40%. In case Win1, in West and East 10% and 17% are saved.
Cases Win2 show savings higher than 16% in all cases except North. Energy
saving with electrochromic windows shows highest values for building that has
glazing on the east side.

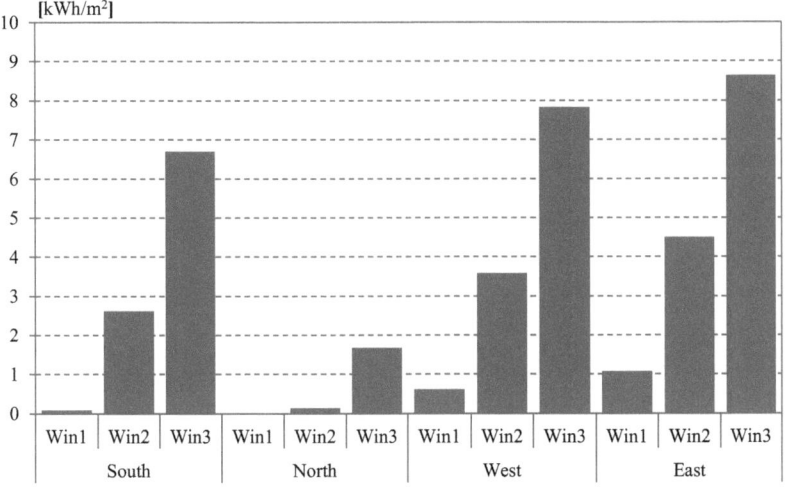

Figure 21   **Reduction in cooling loads per m² - base calculations**

3.3   Illuminance – shading device

Light analysis with shading device works on the same principle as in base calculations. Both types of windows have movable shading devices. In order not to make calculations too complicated, there are only two states of shading device – shading device on and shading device off. Shading device default energy plus shades with medium reflexivity and medium transmittance have been chosen.

Medium Reflectance - Medium Transmittance Shade

Solar Transmittance	dimensionless	0,4
Solar Reflectance	dimensionless	0,5
Visible Transmittance	dimensionless	0,4
Visible Reflectance	dimensionless	0,5
Infrared Hemispherical Emissivity	dimensionless	0,9
Infrared Transmittance	dimensionless	0
Thickness	m	0,005
Conductivity		0,1
Shade to Glass Distance	W/mK	0,05
Top Opening Multiplier	m	0,5
Bottom Opening Multiplier		0,5
Left-Side Opening Multiplier		0,5
Right-Side Opening Multiplier		0,5
Airflow Permeability	dimensionless	0

Table 5   Shade properties

Based on these properties, similar shade was created for Radiance. This object is added to existing building when illuminance is too high. Light calculations are the same for interior and exterior shades, so that distinction has not been made in Radiance (only interior ones have been simulated).

Electrochromic window with shades works in the same way as the electrochromic window without shades for illuminance lower than 750 lux on the measurement point. 750 lux value is taken as the value where still most of the room work surface can be light by daylight. For larger values, illuminance is considered too high, so some sun protection is necessary. When measured illuminance value in the state of lowest transmittance of glazing gets to that value, then shade is deployed. It reduces significantly illuminance as soon as it is deployed. To demonstrate its effect, graphs are presented with electrochromic and normal glazing for comparison.

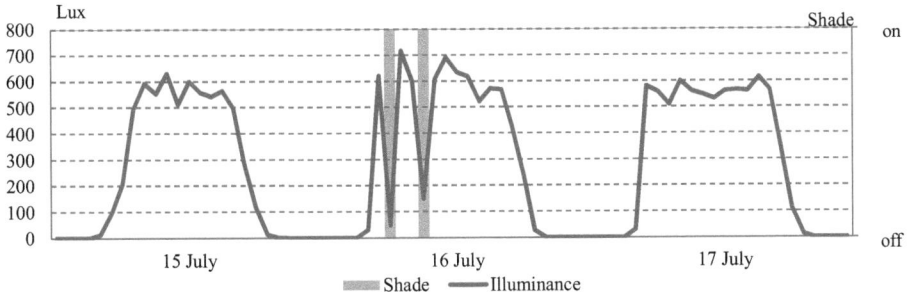

**Figure 22   Illuminance in measurement point, EastIntShade EC
window and Shades state**

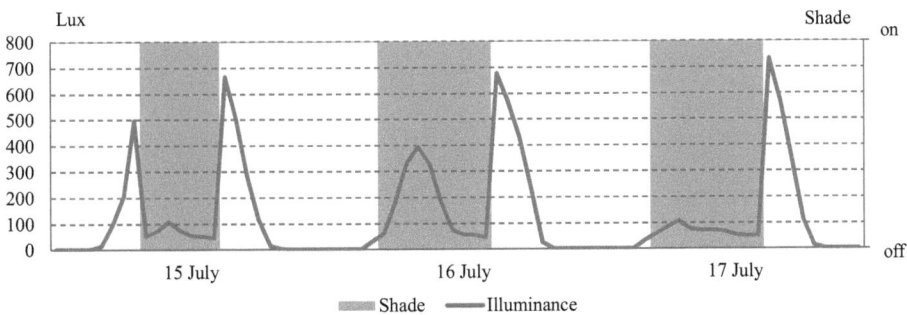

**Figure 23   Illuminance in measurement point, EastIntShade
Normal window and Shades state**

Shades on conventional windows are deployed when same measurement point illuminance value is reached - 750 lux. Because glazing does not change its transmittance properties, shades are deployed more often, and illuminance in the room has more often values lower than 500 lux. Behavior of illuminance on the measurement point is represented for all 4 orientations.

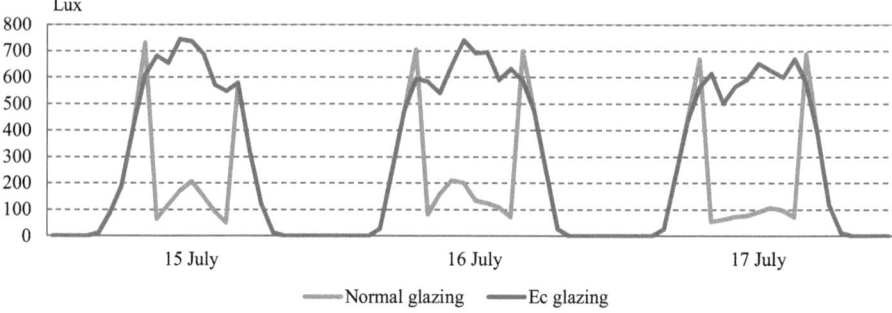

Figure 22   **Illuminance in measurement point for EC and
Normal glazing, case SouthWin3**

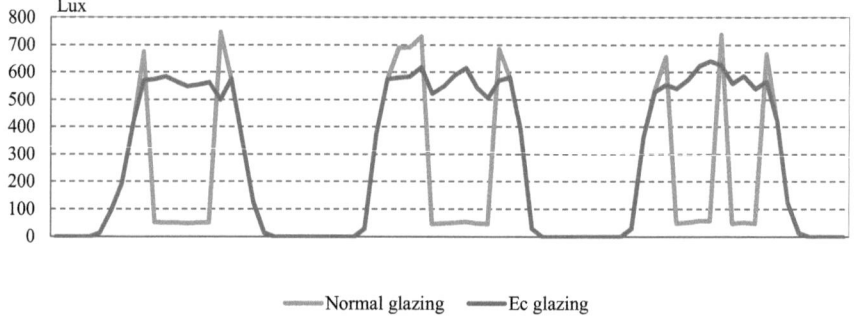

Figure 23   **Illuminance in measurement point for EC and Normal
glazing, case NorthWin3**

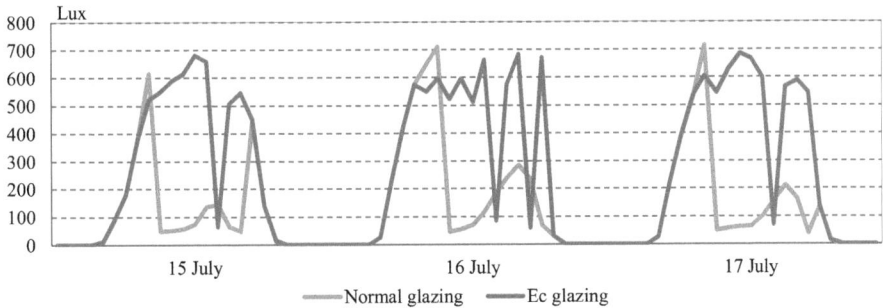

Figure 24   **Illuminance in measurement point for EC and
Normal glazing, case WestWin3**

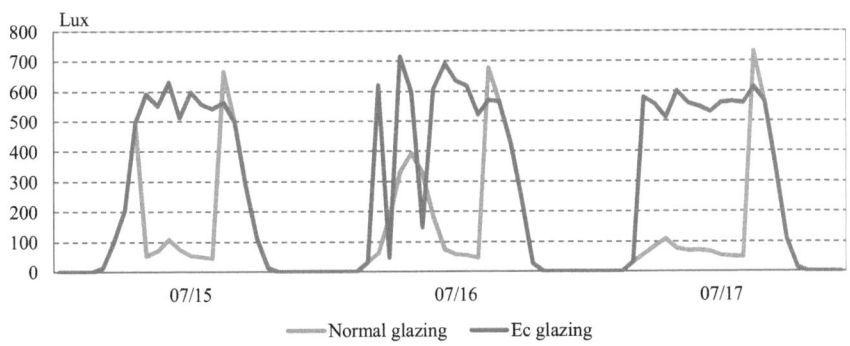

Figure 25   **Illuminance in measurement point for EC and
Normal glazing, case EastWin3**

In shading device calculations, a possibility of using daylight in the room is a very important result beside cooling loads. When illuminance in measurement point gets under 500 lux, it is assumed that whole room doesn't have enough daylight. Because of that, at least in some part of the room artificial light will have to be used. Comparison of daylight hours during summer is calculated when number of hours with illuminance higher than 500 lux is summed up from the performed calculations.

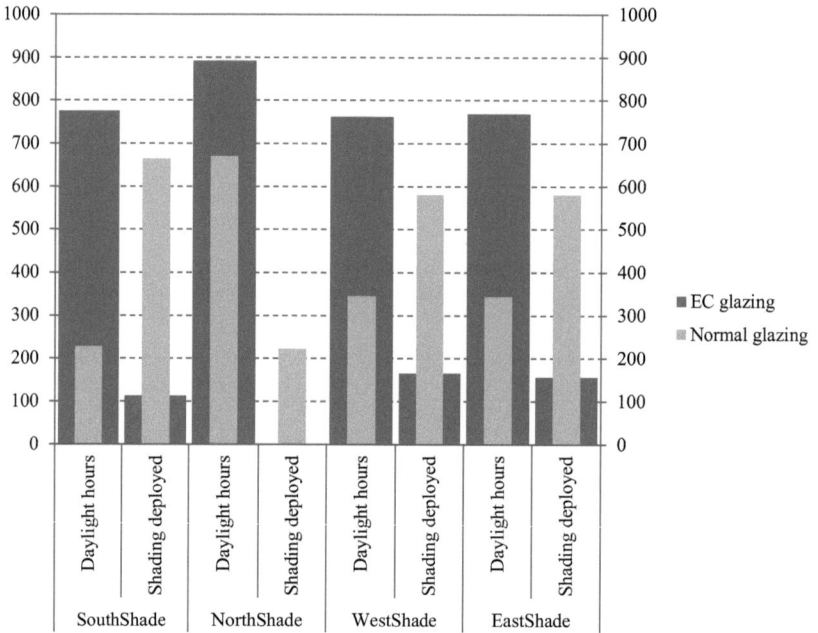

Figure 26   **Number of hours with illuminance higher than 500
lux and Number of hours when shading is deployed**

From the Figure 26 relation between number of daylight hours and number of hours when shading is deployed can be easily noticed. Although the shades have certain transmittance, they don't transmit enough daylight into the room. It can also be noticed that number of hours when daylight provides sufficient illuminance in the room for electrochromic windows is highest on the north. In that case shades are completely unnecessary. Other orientations have approximately same number of daylight hours, but shades are least used on the south.

When normal glazing is used, case with south oriented glazing showed worst results in number of daylight hours and the biggest number of hours with shades deployed. North orientation also shows significant reduction of daylight hours, but still gives best results.

3.4   Energy analysis – shading device

When shading devices are used, cooling loads of building during summer change in different ways. With external blinds, cooling loads are reduced with electrochromic windows, but there is very high cooling load reduction in cases of conventional windows. Interior blinds reduce cooling loads much less than the exterior ones, and in the case of electrochromic windows they even increase it a bit in East and West.

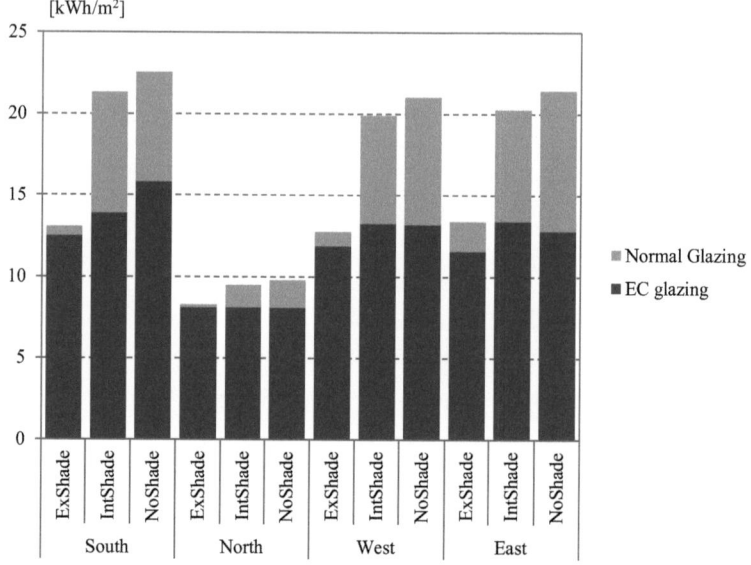

Figure 27   Cooling loads per m² - cases with shading

When exterior blinds are used, cooling energy difference is reduced greatly, but these results can be questioned because of the neglected factors that have been mentioned in the analysis of light. Artificial light that has to be used when shades are deployed increases internal gains, and also cooling loads, besides its regular energy consumption.

Case of cooling load being increased when interior shades are used is going to be analyzed more thoroughly. When electrochromic windows with and without shade are compared, it can be noticed that cooling loads with shading deployed become higher in the peak of cooling loads. In the case of electrochromic windows shades are deployed when illuminance is greatest, and that is usually the time with the peak cooling load. There is some reduction in energy demand when internal shades are used during the whole day. When they are used mostly in the hours of peak cooling load, and when the cooling loads are summed up, it ends up with higher total cooling energy demand for summer period.

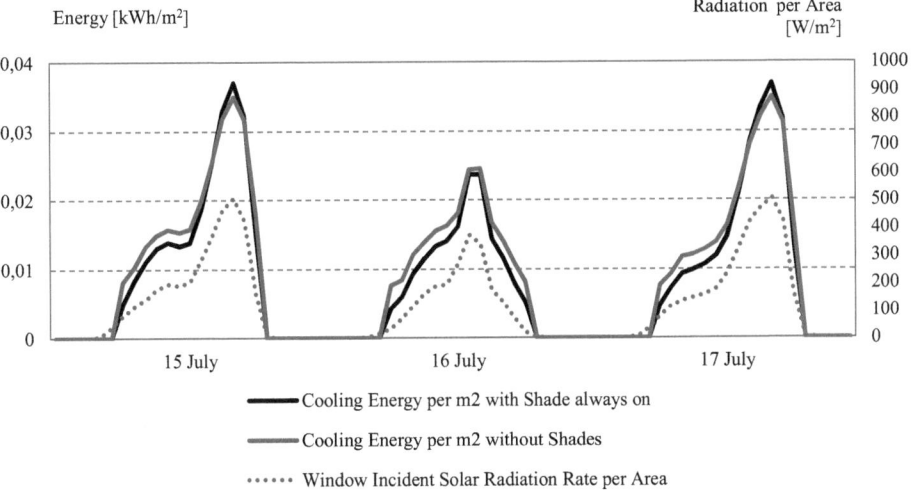

Figure 28   Cooling load per m² - WestIntShade Normal glazing

Why does this happen? "The hourly peak solar load component for a window assembly is affected not only by the use of shades but also by the heat absorbing characteristics of the shade itself. When compared to light colored shades or no shades, dark colored shades absorb more solar heat and release it sooner where it becomes an immediate cooling load in the space. Lighter colored shades reflect a portion of the solar gain back out of the window and a smaller peak load results. Regardless of whether light or dark internal shades are used, the total solar heat load through the window assembly over a 24 hour period is less for windows with any type of shade than without." (United Technologies Corporation. 2006)

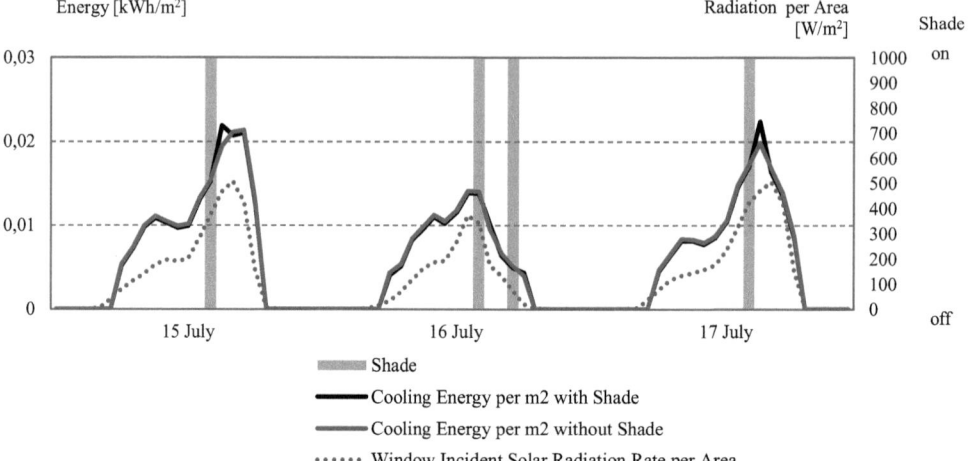

Figure 29   **Cooling load per m² - WestIntShade EC window and
shade deployment**

If the shading deployment is scheduled as in this research, shades are put on only when illuminance is too high. That period differs for electrochromic and conventional glazing. In the case of electrochromic windows, that is also the time when cooling load reaches its peak. If the shades are used only during that time, then they don't reduce cooling loads, but on the contrary, they increase it. This effect can be seen especially in East and West because peaks are more extreme than in the North and South. Because of that total summer cooling load becomes larger than without shades. When normal glazing is considered, shades are deployed much more often than in electrochromic glazing during the day. They are also deployed during the time of the day when cooling load is decreased by their implementation. In this case total energy consumption for cooling during the summer decreases.

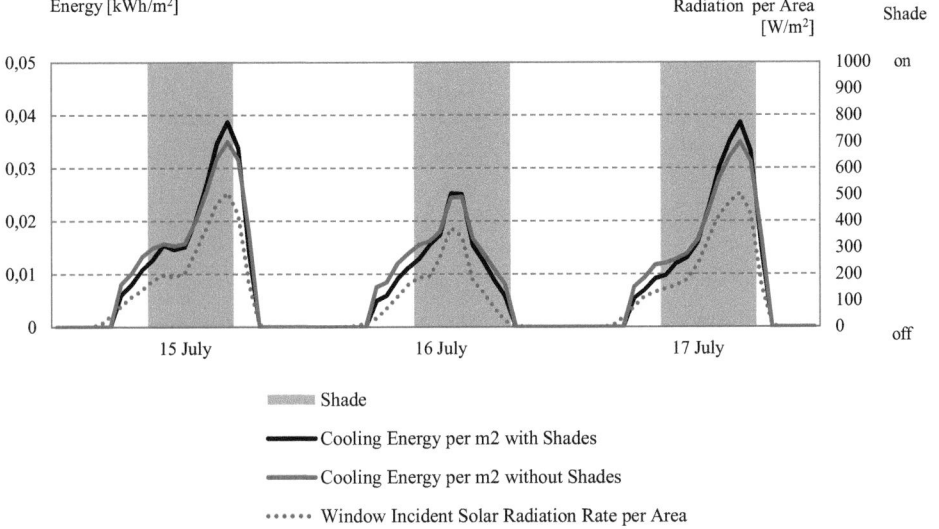

**Figure 30   Cooling load per m² - WestIntShade Normal window
and shade deployment**

In the case of external shades this effect is not found. Energy consumption
for cooling is lower all the time when external shades are used, than when they are
not used. Difference between electrochromic glazing and normal glazing is that
shades are used more often with normal glazing, so graphs differ more with and
without shades, than in the case of electrochromic windows.

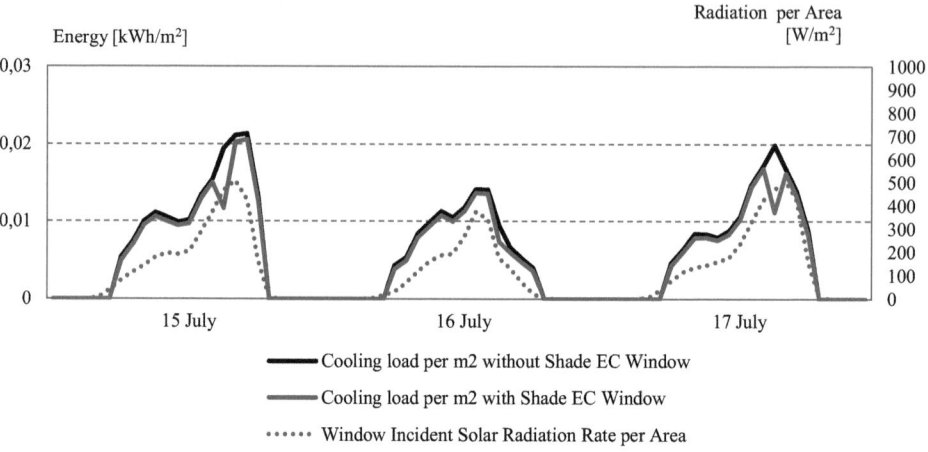

Figure 31   Cooling load per m² - WestExtShade EC window

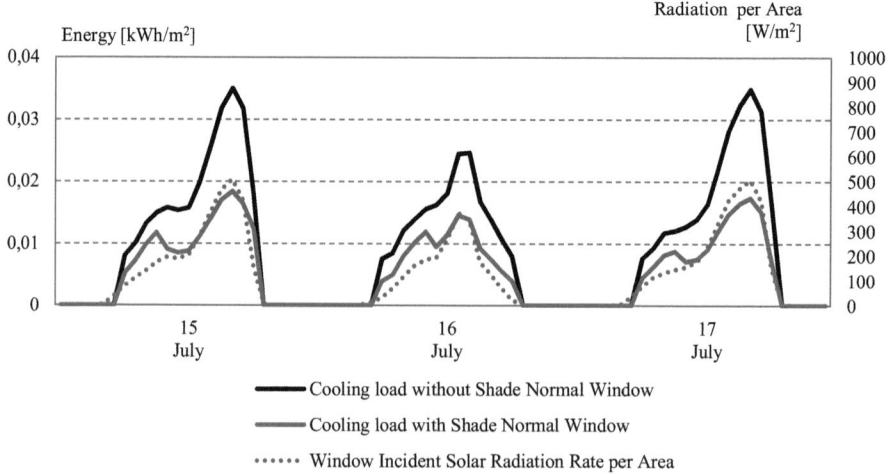

**Figure 32   Cooling load per m² - WestExtShade Normal
Window**

4 CONCLUSION

4.1 Discussion

After this research is performed, there were several important aspects that should be emphasized. First one is related to the process of making simulation with one unconventional building element which is already commercially available. In order to perform simulation with an automated building control, person that does the simulation has to be familiar with, in this case, two simulation programs, one program that couples them to work together, and two scripting languages. Besides that, other software was used for 3d modeling, plug-ins for exporting into simulation programs and software for defining examined building element relevantly. There was no simulation program (up to this point) that could provide possibility to perform easily simulations with building elements that work on this principle. In order to increase the use of new materials, simulation programs should keep up with the development of these technologies, so that the benefits could be proven and calculated by common user.

Another issue that considers the use of smart glazing is insufficient documentation. Products that are currently commercially available don't offer many data about its performance, although they are already used. There are European norms for smart glazing products, but the field is still quite new.

Analysis of cooling loads when smart glazing is used in Vienna climate produced very affirmative and encouraging results. Cooling load has been reduced even 40 % in certain cases. In order to improve thermal comfort and provide more hours with daylight, and retain clear view to the occupants of the building, electrochromic windows can be a very good solution. There are many reasons why they are not widely used, but probably the most important one is lack of information about their use.

Simulation results showed that use of electrochromic windows depends greatly on the size of the windows and orientation of glazing. It is found that in building with glazing oriented towards North electrochromic devices should be used only for large glazing areas. For other orientations they show improvement in energy performance even for smaller amount of transparent surfaces. They can also increase number of hours when daylight is used, and in that way improve conditions for occupants.

This research should provide better insight in the field of smart materials in general, and provoke some other researches to try to precisely determine all the benefits that can be found in these products. Many fields influence development of this technology and some were addressed before.

4.2   Next steps

It is clear that these simulations and the results they have provided do not show exact energy saving potential. However, it is possible to get insight in the relationship of energy demands and smart window devices. In order to get more relevant results few things could be done.

First of all, smart windows used in this research are "active" ones – so they require some amount of energy for their performance. This demand of smart windows is neglected although it produces additional costs. Besides the energy demand analysis, cost analysis could be performed to find out in what time the investment could be repaid.

Building used for simulation has very simple geometry and it is used to find the relationship between smart windows, light and energy saving for Vienna weather conditions. It has no surrounding elements and furniture inside it. For results that would be more similar to a real case improvements could be done on building model. First of all improvements on physical barriers like frames, all kinds of shades and obstacles that affect light which gets to windows should be made. Furniture could also make some change because of its reflecting surfaces.

For cooling load calculations in EnergyPlus, some more complex building with multiple zones, interior gains and other more detailed calculation can be performed. Also, research can be performed for the whole year to find out the benefits and flaws of electrochromic glazing in other periods of the year.

Another important thing would be to compare simulated data with real automated building with smart glazing. As these devices are still not represented adequately in simulation programs, it is not certain how well do the simulations respond to the real case.

There are also other factors which could affect window performance like slow change of phases in cases of some smart windows. Chosen window changes state from dark to bleached state in about 15 minutes, what doesn't affect greatly calculations with timestep of one hour.

4.3   Contribution

After the research of the existing documentation and works done on the topic of electrochromic windows, it was possible to get an insight that researched technology is one of the innovations with greatest potential for future implementation. Along with that, in most of the papers that have been read, lack of research, measurements, and other information has been emphasized.

Simulation software is limited and provides users only with options that producer found to be necessary. It does not exactly follow technological development. For that reason it was not possible to regularly implement electrochromic devices in these programs. Flexibility of the software was the most important factor for performing the simulation. The deficiency of simulation programs regarding this topic has been outdone with the use of additional programs. Documentation, examples and support can hardly be found as they are still not widely used.

This research should close complex simulation performance to the reader, and provide some insight in dealing with uncommon aspects of simulation programs. It can serve as an example of using Radiance and EnergyPlus with BCVTB (coupling of these two programs was not found so far). This research should also emphasize inadequacies of programs for certain new technology and speed up their development.

Results of the research show possibilities of energy savings with this technology in Vienna. It can be related to other areas with similar climate. It should decrease deficit of research data on this topic and provoke further research because of its promising results. Improvements mentioned already can be made, so more accurate results would be found.

In the end, in this time when energy saving is one of the world's most important topics; cooling loads of residential buildings became a progressively popular problem. This technology shows significant possibility of reducing cooling loads, property that might show as revolutionary in the future. Its deeper understanding and wider research would surely provide better climate for its more frequent implementation.

5 REFERENCES

5.1 Papers

[1]   Houghton Mifflin Company. 2006. *The American Heritage Dictionary of the English Language, Fourth Edition.* Boston: Houghton Mifflin Company

[2]   Regents of the University of Minnesota, Twin Cities Campus, College of Design, Center for Sustainable Building Research. 2011. *Windows for high performance commercial buildings.* Internet. Available from http://www.commercialwindows.org/; accessed 10 September 2013.

[3]   Österreichisches Institut für Bautechnik. 2007. *Leitfaden Energietechnisches Verhalten von Gebäuden (Version 2.6)*

[4]   Österreichisches Institut für Bautechnik. 2007. *OiB Richtlinie 6, Energieeinsparung und Wärmeschutz*

[5]   Österreichisches Institut für Bautechnik. 2011. *OiB Richtlinie 6, Energieeinsparung und Wärmeschutz*

[6]   R. McCluney. 1996. *Introduction to Radiometry and Photometry.* Norwood, Massachusetts: Artech House

[7]   National Aeronautics and Space Administration (NASA). Internet. Available from: http://quest.arc.nasa.gov/aero/virtual/demo/research/youDecide/ /smartMaterials.html; accessed 05 October 2013.

[8]   Kroschwitz, J. I. 1992. *Encyclopedia of Chemical Technology.* New York: John Wiley & Sons.

[9]   Addington, M., and Schodek, D. 2005. Smart Materials and New Technologies: For the architecture and design professions. Oxford: Architectural Press.

[10] Baetens, R., Jelle, B.P. and Gustavsen, A. 2010. Properties, requirements and possibilities of smart windows for dynamic daylight and solar energy control in buildings: A state-of-the-art review. *Solar Energy Materials & Solar Cells* 94 (2010): 87-105.

[11] Econtrol. *Dimmbares Sonnenschutzglas: Technische Daten.* Available from: http://www.econtrol-glas.de/econtrol-glas/technische-daten;   accessed   20 September 2013.

[12] Lawrence Berkley National Laboratory, Eleanor Lee. 2006. *A Design Guide For Early-Market Electrochromic Windows* for California Energy Commission: *Public Interest Energy Research Program*

[13] Statistik Austria. 2013. *Energiestatistik: Energiebilanzen Österreich 1970 bis 2011.*

[14] The Board of Trustees of the University of Illinois and the Regents of the University of California. 2013. *External Interface(s) Application Guide.* Internet. Available   from: http://apps1.eere.energy.gov/buildings/energyplus/ /energyplus_documentation.cfm; accessed 15 September 2013.

[15] Jacobs A. 2012. *Radiance Tutorial.* Internet. Available   from: http://www.radiance-online.org/learning/tutorials; accessed 10 May 2013.

[16] *BCVTB   official   website.* Internet. Available   from: http://simulationresearch.lbl.gov/bcvtb; accessed 10. September 2013.

[17] IBO - Österreichisches Institut für Baubiologie und –ökologie. 2009. *Passivhaus-Bauteilkataloges.* Available from http://www.baubook.at/phbtk/index_BTR.php?SW=19; accessed 2 September 2013.

[18] Kurt Battisti. 2008-2013. *Die Bauphysik-Datenbank.* Available from http://www.bphdb.com/; accessed 2 September 2013.

[19] International Glazing Database (IGDB). Internet. Available from: http://windows.lbl.gov/materials/IGDB/; accessed 20 August 2013.

[20] United Technologies Corporation. 2006. *HAP e-Help 009 (Rev 1)*. Internet. Available from: http://www.docs.hvacpartners.com/idc/groups/public/documents/ /marketing/hap_ehelp_009.pdf; accessed 10.10.2013.

5.2   Software

[21] The Regents of the University of California, through Lawrence Berkeley National Laboratory (2008 – 2013). *Building Controls Virtual Test Bed (BCVTB) Edition (Version 1.3.0);*
Available from http://simulationresearch.lbl.gov/bcvtb/Download

[22] The Regents of the University of California (1995-2011). *Ptolemy II (version 9.1.devel);*
Available from http://ptolemy.eecs.berkeley.edu/ptolemyII/ptII8.0/download.htm

[23] The Regents of the University of California, through Lawrence Berkeley National Laboratory (1990-2002). *Radiance (Version 4.2.a-Linux);*
Available from http://radsite.lbl.gov/radiance/framed.html

[24] US Department of Energy (2013). *Energy plus (Version 8.0.0);*
Available   from
http://apps1.eere.energy.gov/buildings/energyplus/energyplus_download.cfm

[25] The Regents of the University of California (1999-2013). *Window (Version 7.1.73.0);*
Available from http://windows.lbl.gov/software/window/7/index.html